I0503920

1926

AMERICAN SCENES

GENE ANDREW

Jung Works

Copyright © 2023 by Christine Jung.

All rights reserved.

No part of this work may be reproduced, stored
in a retrieval system or transmitted in any form
by any means, electronic, mechanical, photocopying,
recording, or otherwise, without written permission
of the publisher. For rights and permissions, please
contact:

Jung Works

christine@jungworks.com

Book Cover and Illustrations by Christine Jung

1st Edition 2023

CONTENTS

INTRODUCTION

To me, Gene Andrew was "Granddad." He was gentle but strong, oh so smart, and had a great laugh and sense of humor. I have many wonderful memories of my own with Gene: checking on the newborn calves, taking a walk through his garden (is there anything more rewarding than to see your plants grow and getting to pick fresh, ripe tomatoes?!), and making home-made lemonade and ice cream.

But above all, I remember Gene to be a storyteller. He wasn't a writer. No, if he was putting pen to paper, it was for accounting purposes (he was meticulous with his farming and ranching records) or to solve the newspaper's crossword puzzle of the day.

Rather, Gene's art lay in verbally relating his stories. How he could spin a tale! His adventures came to life in the telling. They were all true stories, but the way he told them – I was right there re-living them with him.

Sadly, I have forgotten so many of the stories he told me over the years, so I am ever so thankful I was able to at least squeeze these writings out of him (he was a reluctant writer, but I didn't let up!) He only got so far before he fell ill with, and ultimately succumbed to, cancer.

I am honored to share Gene's memoirs with you.

I hope you are entertained by them. Perhaps you will learn something new or take a walk down memory lane yourself. I know Gene would be proud to know his stories live on.

-Christine Jung

(Gene's Granddaughter)

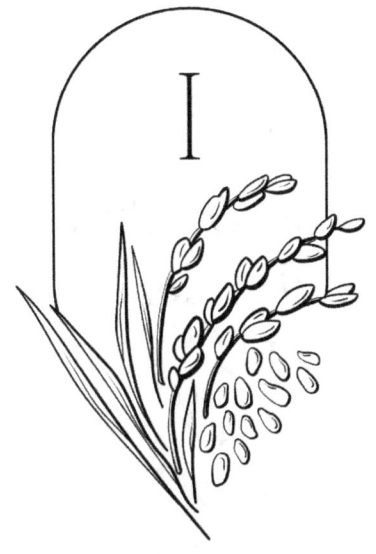

IN THE BEGINNING

I was born November 16, 1926, near the small town of Grand River, Iowa. My sister, Doris, was born April 7, 1918, my brother, Harold, March 25, 1924, and my other brother, Wayne, December 6, 1931. We were all born in the same house,

probably in the same bedroom, located
on a black-land farm in Southern Iowa,
about five miles south of Grand River, on
a narrow dirt road.

*Editor's Note: Grand River was founded
in 1881. The 2020 census counted a
population of 196.*

Gene Andrew 1947

Grand River, Iowa

We were delivered by a country doctor, I don't recall his name, who made house calls in his Model T Ford, and perhaps in horse and buggy before that. Usually, the first order of business by the doctor was "get me a kettle of hot water." The dirt roads being sticky black mud when wet made travel very difficult during rainy weather, and sometimes it took the doctor several hours to reach a remote farmhouse, many times at night.

Usually, the first order of
business by the doctor was "get
me a kettle of hot water."

Our one hundred sixty-acre farm was
located in Decatur County, about twenty
miles north of the Missouri state line. This
was very near the Mormon Trail on which
Brigham Young led his followers from
Eastern Illinois to Utah. We were located
near the center of a circle of small towns
– Grand River, Decatur, Leon, Lamoni,
Tuskeego, Bloomington, and others.

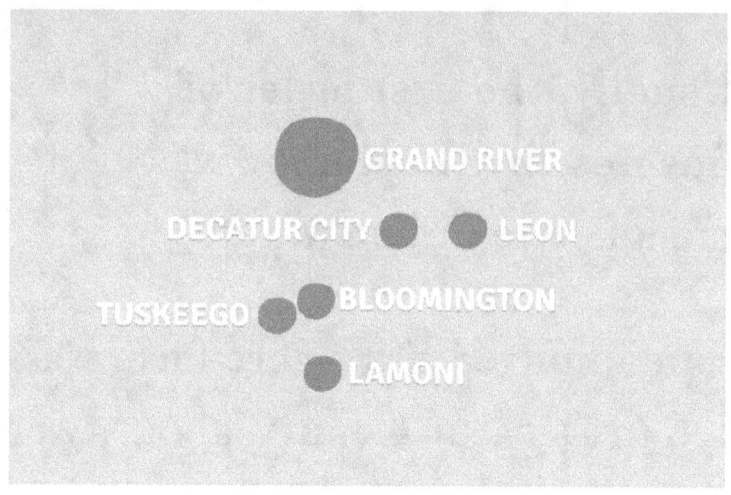

Grand River and the Surrounding Towns

A BRIEF FAMILY HISTORY

My father was Lyman Irl Andrew, born October 16, 1890, and died September 2, 1972, at the age of 82, and was the eighth of twelve children:

six brothers and five sisters. The brothers were Jesse, Allen, Lewis, Harry, Arthur, and Paul. The sisters were Cora, Grace, Verna, Velva, and Vesta.

Andrew Home in Iowa

Uncle Jesse died before I was born at the age of thirty-six, but all the other uncles lived in our immediate area. I idolized these uncles, and my view of them was that they were really great men, in more ways than one, and that they could do no wrong. I relished the times I could go with Dad to visit them and my cousins. I believed they helped shape my life in many ways as they were a great influence on me.

For instance, I never heard any of them use a word of profanity. The Andrew brothers were big men physically. Dad always said he was the "runt" of the family, and he weighed 198 pounds in his prime. They had a reputation far and wide for their physical prowess and seldom lost an opportunity to display it.

Lyman and Myrtle Andrew 1937

Dad dropped out of school after the fourth grade, as did most others of that day, to work on the farm and help earn a living for the family. Despite this, he was a self-educated man and had a great knowledge that few possess today.

Dad dropped out of school after the fourth grade, as did most others of that day.

Dad was a hard worker and was not afraid of hard manual labor. Many days, he would come in from work completely wet with sweat from head to toe. There was very little labor-saving equipment, and he did everything by hand to eek out a meager living for the family. He could build a quarter mile of fence in one day and every post was dug with a hand digger.

In the year 1910, when Dad was nineteen years old, one of his friends became ill with some kind of respiratory ailment. The doctor recommended that he move to an ocean climate for his health, so the Texas Gulf Coast, being the nearest, was chosen.

This friend had kinfolks in a little town called Katy, so he asked Dad to accompany him to Texas.

His friend became homesick and returned to Iowa, but Dad liked Katy and Texas so much, he stayed two years. He worked for many of the rice farmers of that time and was acquainted with many people in the Katy area. One winter he worked in a rice mill in Katy, known recently as Katy Warehouse, and has since been torn down. Later, he worked a few months on the Galveston Causeway, a long bridge from Galveston to the mainland. He then returned to Iowa in 1912.

Editor's Note: In the early 1900s, Texas and Louisiana together produced 99% of the United States' rice.

Lyman Andrew in Katy, Texas 1910

I suppose the fact Dad came to Texas was sort of a pre-destination for me and the rest of the family. This event was to drastically change my life, and it took place before l was born.

Upon his return to Iowa, still having a roving and adventurous spirit, he then went to Canada for about two years where he worked on a wheat farm growing many hundreds of acres of wheat with horses and mules.

About the time he left for Canada, one of the one-room schools in the area, in Tuskego, hired a young schoolmarm named Myrtle Brown. It so happened that Uncle Roy Brown and Aunt Grace lived in this community, and one day when Dad went to visit them, he saw Myrtle and asked, "Who is that pretty girl?" Of course, things progressed, and they later corresponded after he went to Canada. On his return, they eloped to Des Moines and were married on May 6, 1915.

I did not know my aunts, with the exception of Grace, whom I faintly remember. One of them, Velva, died in infancy and the others married and moved away. Aunt Grace married Mama's brother, Roy, so I have five double cousins: Enid, Revere, Virgil, Bruce, and Earl Brown.

They moved to Idaho when I was seven or eight years old, and I never saw them again after that. I'm sure their families are scattered around Idaho and the West Coast. One incident concerning Aunt Verna made the headlines in November 1950. Her son, James Andrew Huff, was killed in Montana in the crash of a commercial airliner that he was piloting.

My mother was Myrtle Bertha Brown Andrew, born on May 23, 1893, and died October 22, 1986, at the age of 93. She was the sixth of eight children: three brothers and four sisters. They were Minnie, Fred, Helen, Horace, Roy, Hazel, and Temple. Mama was born and raised at Tuskego near Lamoni in Decatur County. I have no knowledge of where she attended school except that she went to Graceland

College, a small college founded by the Mormons, as was the town of Lamoni. Graceland College's best-known alumnus is the great athlete, Bruce Jenner, who won the decathlon event in a recent Olympics.

She went to Graceland College, a small college founded by the Mormons.

One of Mama's sisters, Minnie, went along to Graceland to do the cooking, ironing, etc. – sort of a private maid. I have very little recollection on the Brown side as they lived farther away, but we did visit some of them occasionally.

My favorite was Aunt Hazel and her husband Clyde Hembry. It was a treat to go to their place, and we had great times with cousins Mary Belle, Myron, and Don. Later, there were two more girls: Joanne and Carol Jean. Dad and Uncle Clyde always kidded a lot and I believe dad was in his best humor when he was around Uncle Clyde. Uncle Clyde would not argue politics or anything else, he would just nod his head and say, "Yeah, yeah, Lyman".

Aunt Temple Cooper lived in Denver and for many years always sent a Christmas package which we eagerly looked forward to.

Aunt Minnie was an old maid and lived with Grandma until she died. I remember her best for her little deformed left hand which she was very self conscious about. We kids always wanted to see her peel potatoes and use her "little hand".

My grandparents and great-grandparents were some of the early settlers and pioneers of the Midwest.

My grandparents and great-grandparents were some of the early settlers and pioneers of the Midwest. Grandpa John Newton Andrew and Grandma Clarissa Arabella Brown I knew very well as they lived to old age. In fact, Harold and l witnessed the death of Grandpa Andrew on July 4, 1934. Grandpa Clifton Albert Brown and Grandma Evaline Eldora Akers both died before I was born. One great grandfather, Samuel McDowell, was a stagecoach driver.

Great-grandfather Spencer Collings Akers is a story by himself. He was one of those rugged individuals who pioneered the westward movement in the USA. When he was eighteen years old, he married Elizabeth Ader, fifteen years old, and they then struck out for California and the Gold Rush in 1849 in a covered wagon.

Four of their first five children died in infancy. One was born in a covered wagon. They had fourteen children altogether, and the third surviving child was my grandmother, Evaline Eldora.

In 1852, Spencer Akers returned to Iowa via boat to Panama, where he hired Native Americans to transport his tools and belongings across the Isthmus of Panama on foot, then took another boat to New

Orleans and a steamboat up the Mississippi to Burlington, Iowa.

Being a blacksmith and a carpenter, he took his tools to California. Among the tools he brought back was a hundred-pound anvil, used for blacksmith work. Later he gave it to Dad who gave it to Harold and me to use on the farm. I ended up with it and sold it in a farm auction for fifty dollars. Spencer Akers operated an inn and post office near Elk Chapel, a half mile south of East Elk school. He was called upon to be a doctor many times and did amputations without anesthetics, administered flu serum, etc. His wife, Elizabeth, was a mid-wife, and would go out in the middle of the night, on horseback, in the snow to deliver a baby.

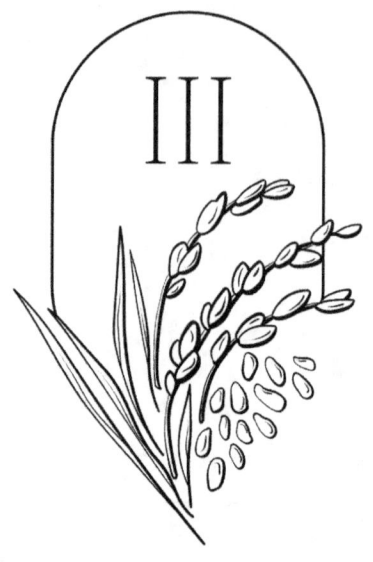

EARLY MEMORIES

My earliest recollection is when I was four years old. I was playing at my mother's apron strings, literally. I started to school at east Elk school, a one room school one and a half miles from home, when I was four years old. Doris, who

was twelve, had finished the eighth grade
and was to go to high school in Lamoni
and room and board with Grandma Brown.
Harold would have had to ride Scotty, our
Shetland pony, by himself. That would
have left a place open on Scotty, so my
first teacher, Madge Woodard said, "Just go
ahead and send Gene too." So, I started to
school and must have done very well since
I progressed and advanced two grades in
some years.

East Elk School Class Photo – 1935:
Harold Andrew – Top Row 2nd from
Right, Gene Andrew – Bottom Row 2nd
from Left

My favorite activity at school was cutting out black cats, leaves, and other things for Halloween. Back then, the winters were very harsh with snow drifting over the tops of the fences and filling the roads. Harold and I rode Scotty to school most of the time, and when the snow filled the roads, crusted over where it would hold up the pony with us riding, we could cut across the fields right over the tops of the fences. When the snow was fresh, we had to walk to school. The only time we stayed home was if there was a blizzard or if we were sick. I don't know how Mama could let us go to school under those conditions, but we thought nothing of it.

My favorite activity at school was cutting out black cats, leaves, and other things for Halloween.

We rode Scotty bareback with only a bridle; the saddle was too much trouble to put on. Many times, we were dumped in the snow or on hard ground when Scotty shied. This was a jump sideways when a bird or something startled her. We would take a very hard fall, but she would always come back and stand there waiting for us to mount again.

The one room school had two rows of seat-desks with the old-time ink wells; the small seats in front for the younger ones and gradually larger toward the rear. There was a long wooden seat at the front of the room called the "recitation" seat and a huge blackboard on the front wall. Each class would come forward at their turn and sit in the recitation seat.

My next teacher was Dena Burke, and I believe these teachers taught patriotism, country, and God the way it should be taught. We said the Pledge of Allegiance and sang America the Beautiful every morning, then raised the flag.

I was privileged to visit with Madge Woodard in Leon a few years ago. She was in her late eighties and was very active in city politics. She could remember which seats we had at East Elk and could remember about all her students in the other schools.

East Elk School, Decatur County, Iowa

At East Elk, there was a man-made storm cave about a hundred feet west of the school house for emergency use in the event of a tornado. It was never needed. Harold and a classmate, Harold Akers, had a fight down in this cave. They were locked in a deathly struggle. They were pulling hair and squeezing each other, and l was really scared, but someone got the teacher, and the fight was broken up. They were good friends afterwards. A few years ago, we visited the old school, and all the old desks were down in the cave. I put one in the back of the car and brought it home for an antique.

One of my favorite times was recess, and another was the hour-long noontime. The recess was fifteen minutes, and many times we would play tag. There was one boy faster than I, and he could always catch me.

Most of the children rode horses to school and they would take their saddles off and tie the horses in a group at the south fence of the school yard. There was no shed, and many times in the winter the horses would be covered with snow or sleet when it was time to go home.

Our route to school was a quarter mile south, a mile east, and another quarter mile south. At the first corner, we were joined by Margaret and Junior Rauch. Junior always had a plug of tobacco or a bottle of wine he had sneaked from his Uncle Halley. Halley and his father, Jim, lived about halfway down the road to school and were bachelors. I was always relieved to get past there as l was afraid of them. Their main occupation was making homemade booze and consuming it right there at home.

One day, on the way home from school, Junior gave Harold a chew of tobacco. Harold quickly became very sick and dizzy. He could not continue on home, so he climbed a steep road bank and laid down under a hedge tree near the fence. I continued on home and told Dad where Harold was lying. He took the mules and wagon and went and got him. Another time, Junior had a bottle of wine which he drove a nail through the bottle cap and made a small hole to sip through. He hid the bottle under a bush and every day on the way to and from school, he would take a little swig and then put the bottle back. He made it last a month or so.

One year, at the end of the school term, we were to have the school picture made. Our cousin, Ray Andrew, was into photography and was sort of the community photographer. He came that day to take the picture, but I got really stubborn and refused to be in that picture. The teacher tried everything, including begging and threatening me, but I never did get in the picture, and I don't know why.

As mentioned before, about the time I started to school, Doris was ready to go to high school and was to room and board with Grandma Brown in Lamoni. Now Doris is a very talented artist — especially drawings and sketches. She could have been a professional cartoonist or fashion artist. One day, she idly drew an uncomplimentary cartoon of Grandma. It wasn't meant for her to see the cartoon, but she found it and their relationship deteriorated after that.

The next year, she went to Grand River High School. She lived at home and rode Scotty north five miles to Grand River, rain or shine, sleet or snow. She was joined on the way by cousins Ralph and Vernon on their horses and further along by others. Harold and I walked to school that year as Doris was riding Scotty.

Then she attended Leon High School, a town fourteen miles east, where she had room and board. Some weekends, Dad would go get her in the car. One weekend in a terrible blizzard, Mama was real worried because they were so late getting home.

Doris graduated from Leon High School in 1936. My younger brother, Wayne, started to school that year and was in the first grade when we moved to Texas.

I believe I received a very good basic education in this one-room school. The building is still standing at this time and is being used as a hay barn by a farmer. I have a picture of myself standing beside it taken about 1984.

Gene (L) and Harold (R) Andrew –
1937

THE GREAT
DEPRESSION

I believe the Great Depression was one of the three most important historical events of my life, the other two being World War II and the landing of a man on the moon – which ranks near the top of the list of great accomplishments.

> **I think we were almost unaware that the depression was even happening. We grew up in extreme poverty.**

I think we were almost unaware that the depression was even happening. We grew up in extreme poverty and I would say there were other folks worse off than we. I would describe their circumstances as grinding poverty. There was no welfare, no jobs, and no money. Nearly all the cash had disappeared and gone out of existence, so if you had anything to sell, there was no one with any money to buy it. But there was no one, on radio or TV, telling us how bad off we were, so everything seemed only a normal condition.

But there was no one, on radio or TV, telling us how bad off we were, so everything seemed only a normal condition.

We did have plenty to eat and never went hungry. My Dad had planted a large orchard of apples, plums, pears, peaches, grapes, and mulberries when they first married. The trees grew fast and were large enough to bear fruit in just a few years. They bore many times more fruit than we could eat. There was the flock of chickens for eggs, three or four Jersey cows for milk and butter, and hogs to butcher about twice a year. Also, there was a large garden.

Butchering a hog was an experience. Dad would pick a cold, frosty morning, build a big fire under a barrel of water, and when the water was scalding hot, went to the house for his old single shot .22 rifle and shot the pig between the eyes. The pig's hind legs were fastened to a single tree, part of a harness set, and was hoisted up by means of a small block and tackle fastened to a sturdy tree limb. Then it was lowered into the scalding water for about three minutes to loosen the hair. It was our job to help scrape the hair off. It came off easy but every square inch of the body had to be gone over with a scraping tool. If the hair wouldn't slip off easily the pig had to be re-dipped in the hot water. Then Dad proceeded to butcher the pig, quarter it, and cut up the pork chops, shoulders, ham, and bacon. The bacon sides were put into the smokehouse for a few weeks to make bacon. A large portion was ground into sausage, probably making about fifty pounds. Mama canned most of this sausage in Mason jars and also some of the pork chops. So, we had pork chops the year 'round, but I cannot remember ever butchering a cow.

The fat was rendered into lard in a large black pot in the front yard. Then it was made into homemade soap by adding lye and other ingredients. The soap was used entirely in the washing machine.

Here I have to digress a little and tell an old-time joke about hog butchering. In the old days, it was a common practice to borrow meat from your neighbor if you were low on meat. Then, when you butchered, you paid them back. It was borrowed piece by piece such as a shoulder, or ham.

One day two neighbors were talking. One said, "You know, it is about my turn to butcher a hog, and I have borrowed so much meat that when I pay everyone back, I won't have any left for myself." The other one said, "I have the solution. You go ahead and butcher out in plain view and hang it in the tree where everyone can see it. Then when night comes, take it in and hide it and tell everyone that your hog was stolen during the night. Then stick to your story, and don't let anyone make you change it."

The first neighbor thought that was a good idea and planned to do it. The second neighbor had a better idea. After dark, he sneaked over and stole the hog for himself. Upon seeing his neighbor the next day, the hog's owner said, "Someone stole my hog during the night." And the other man replied, "Now be sure and stick to your story."

Mama also canned lots of fruit, vegetables, preserves, pickles, jellies, and anything that could be preserved in Mason jars. There were quarts, pints, and half gallons and were stored in the storm cave as it was cool down there the year round.

Most of the time, we had excess butter and eggs which were taken to Grand River or Leon, maybe twice per month, where they were traded for things we could not produce on the farm such as flour, sugar, shoes, etc. Most people went into town on Saturday, so all the towns were a beehive of activity that day.

One of these trips to Leon, we parked on the street around the courthouse square. The parking was on an angle and we were beside a brand new, shiny '34 Chevrolet. I looked that car over very thoroughly and thought, oh how I wish we could get a new car; maybe someday I would own one. That is where I learned to really appreciate a new automobile. The fact that a few automobiles were being sold shows that some folks had money to spend, but we had little of it.

Editor's Note: Living the American Dream, Gene was, in fact, able to buy his very own new car later in life. It was hard-earned, and well-loved.

Many of the young men joined the C. C. C. (Civilian Conservation Corp), one of F.D.R.'s make-work programs. This program was set up sort of like the army; they even wore uniforms similar to army uniforms. They planted forests, built dams, streets, sidewalks, and any other project found for them. They were very well known for leaning on their shovel handles.

FARM LIFE IN IOWA

Although our farm was small by today's standards, about one hundred sixty acres, it produced plentiful crops and livestock. This was because it was good blackland, and Dad was a hard worker with a natural-born instinct for farming. He farmed a fairly large acreage of corn and oats, with a smaller acreage of timothy and alfalfa hay. He also kept a herd of beef cows, and about eight to ten brood sows which produced about one hundred pigs a year. Nearly the entire corn crop was fed to the hogs as the corn brought more money on the hoof. The oats were for the mules, and the hay for wintering the other livestock. Dad also farmed a tract of land across the road west which had a rich bottom known as the Bullard slough. I believe this land was rented from Uncle Allen.

It was to this creek bottom that Dad sent Harold and I to cultivate a field of corn that had been up for a few weeks and should have been six inches tall. I was probably six years old, but I had to ride on an iron

seat right behind the mules tails and guide
the team while Harold rode the cultivator.
We could not understand why Dad sent
us to cultivate this field, as it was barely
peeping out of the ground. We went ahead
and cultivated it and covered most of it
up. The problem was that Dad had two
separate fields then, and we had gone to
the wrong one, but we knew better than to
come home without finishing the job.

We spent considerable time wading in
this creek catching minnows and crawfish
while Dad worked in the fields nearby.

During this time, my mother was in poor
health and did not feel like trying to keep
up with two young boys, and later a third
one. I think it was because of this that we

were allowed to roam almost at will all over the farm.

I spent my entire first ten years on this farm, never spending a single night anywhere else. We spent a lot of time in the orchard which was just adjacent to the house to the north. It also served as a windbreak in the winter. We knew almost every tree and climbed them to eat any kind of fruit that we wanted. Sometimes we ate apples that were still not ripe and became very sick. We even climbed the mulberry trees and ate mulberries.

I spent my entire first ten years on this farm, never spending a single night anywhere else.

We spent many hours at the corn crib –
a large barn where the corn was stored.
The hogs were pastured near the barn
and were fed corn directly from the stored
corn. Dad shucked the com by hand and
brought it in a wagon, then dumped it into
a homemade elevator hitch that took it up
to the eave and dumped it in through a
hole near where the roof and eave joined
together. A good crop of corn was about
seventy-five bushels per acre – nowadays,
about one hundred fifty bushels.

Andrew Barn Iowa – Built 1934

There was a definite art to shucking corn. First, you had to have a team of horses or mules that would move up on command and stop on command. The shucker was on foot, following the row of corn. The wagon was long, narrow, and deep with a tall sideboard extended on one side to deflect the ears of corn down into the wagon as the shucker threw them. A lot of the farm boys started shucking corn as teenagers. The best ones entered competitions at the various county fairs, then competed for the state championship and then the national championship. The champion was nearly always from Iowa or Illinois and in his early twenties and could shuck one to two hundred bushels per day. It was said that a good shucker could keep an ear of com in the air all the time.

Uncle Harry owned the only threshing machine in the community and did custom threshing with it. It was powered by a huge Rumley tractor, a gasoline tractor, but before that he had a steam engine. Each year, he made the rounds, threshing wheat, oats, barley or whatever. This was a great time of year for us kids since we were too young to work in the fields but got in on the feast that is a tradition for the threshing crews. The crews were made up of neighbor men who came to help and didn't expect any pay. Then Dad would go help them when they threshed their crops. The only crop we grew for threshing was oats.

Dad had a shop where Harold and I spent many hours making things – or tearing them up. The main equipment in this shop was a forge and the old anvil mentioned earlier. This is where Dad sharpened the plowshares and cultivator sweeps and also could weld a little in a crude fashion by heating the metal red-hot and beating it together with a sledgehammer. It was my job to turn the crank for the air blower which quickly heated the coal in the center to a white-hot temperature.

Back then, the winters were very harsh. The snow would drift against the house up to the eaves, and some mornings, it was difficult to get the screen doors open. The snow kept piling up as it did not have time to melt between snowstorms.

One year, a lady died in Grand River, and they could not get the body to the cemetery because of the deep snowdrifts on the road. There must have been over a hundred volunteers to shovel the snow and get the road which went past our house open for the funeral. Some started at the north end, some at the south end, and shoveled toward the middle. Finally, after a few days, the road was open enough to let the hearse through, and then another snowstorm filled the road full again.

We went to school every day and were sick very few times. I believe you build up a resistance to the cold weather.

Most of the time, wood was burned in the heating stove and the cook stove. When we got low on wood, Dad would go to the woods and cut a wagon-load of logs, all by hand with an ax. Then, it was sawed in two with his homemade circular saw. It was an old car frame with the engine on it which powered the thirty-inch circular saw blade. Then part of my chore was to chop or split the short logs with the double-bitted ax.

During the few days the road was open, Dad ordered some coal for the big living room heater. It was delivered in one huge lump by a dump truck that just dumped the load beside the driveway out by the road. It was about six feet long and about three feet thick and must have weighed two tons. We just chipped it off in little pieces as it was needed by hitting it with a sledge and carried it in the house in the coal bucket.

We always had chores to do, both before and after school.

We always had chores to do, both before and after school. They included milking, feeding the chickens, splitting the wood and carrying it in, and sometimes feeding hogs or mules. I did the lighter chores, and it was Harold's job to do the milking, usually two or three Jersey cows, which gave a huge amount of milk. Many times, I would be standing around the barn when he was milking and if I got within about fifteen feet, I would stand still and open my mouth real wide and he would squirt milk in my mouth. He had a very good aim, but it always got all over my face.

The mules were Big Jenny and Little Jenny. One stood about a hand taller and weighed about a hundred pounds more than the other. Dad claimed they were the best team in our part of the state. He did all the heavy farm work with them. They were always ready to go. It took about ten minutes to put the harness on, and many times when the harness was taken off after a hard day's work, they would roll and roll on their backs as it felt so good to them to have the harness off. Then they would get a big drink of water at the stock tank. I always wondered how they could drink so much, but they were only watered twice a day. Most farmers worked horses, but Dad believed that horses were not as strong and did not have the stamina that mules had.

By the 1930s, there were beginning to be a few REA (Rural

Electrification Administration)
lines built, but none came our
direction.

By the 1930s, there were beginning to
be a few REA (Rural Electrification
Administration) lines built, but none
came our direction. So, Dad put in
his own electrical system. It consisted
of a large set of Edison storage
batteries and a one-cylinder engine that
powered a generator. These batteries
were actually a series of several batteries
connected together that would store
enough electricity for a week at a time.

We only ran the washing machine and
electric lights from this supply. When the
lights would start to get dim, it was time
to start the generator and charge them up.

Uncle Harry had a wind charger, but most people just did without and used a kerosene lamp.

Our water supply was from a cistern dug near the top of a gradual sloping hill around a third of a mile to the northeast of our house. This cistern was built like an inverted funnel – about twelve feet deep and twelve feet wide at the bottom. It was lined with brick and mortar and tapered to the top to a small concrete dome with a lid on it. I could never figure out why Dad put the cistern where he did, but I later learned that it was over a spring, and also the water would run by gravity and flow to the house through a pipeline. So, we had running water, something that most people did not have.

There was also a windmill to the northeast a half of a mile, located in the pasture like the hub of a wheel where all the pasture fences came together and all the cattle could water from one source. There was also a windmill west down by the Bullard slough, and when the stock tanks were getting low on water Dad would send us, usually on foot, to tum the windmill on. And a little later, "Go turn the windmill off." One pasture "Over North," as it was called, had its own hand pump and about a two-hundred-gallon oblong tank. Harold and I would have to go every few days and pump this tank full.

The summers were nearly always hot and dry. There were many days over one hundred degrees. Occasionally, there would be a very violent storm, always at night. Just after dark the lightning would begin in the far northwest. It would get closer and closer, and sometime after midnight, it would hit with lightning, thunder, heavy rain, and a furious wind. Just before it struck, Dad would get us all out of bed and make a mad dash to the cave – a homemade cave with sloping doors at the entrance over a set of concrete steps going down. It was covered over with a foot or so of dirt and served two purposes: protection from tornadoes and to store the milk, vegetables, fruit, and canned goods. When we ran to the cave, we grabbed blankets and took them along, and Mama wrapped the baby, Wayne, in a blanket and ran with him in her arms. The storm would rage for a half an hour or so and Dad would occasionally open one of the sloping doors and look out to see if there were any tornadoes. The big maple trees around our house would bend nearly to the ground in the terrible wind. Finally, it would pass and afterwards, it would be just as calm as could be. We would go back to the house and go back to bed. We never did have a tornado that I remember.

I could write about many more things, but this chapter must close somewhere. I think my first ten years were about as eventful as any kid has ever had, and those years provided me with a very good foundation that few people are privileged to have.

The Dust Bowl

We did not realize how severe it was until later.

In the summer of 1934, a severe drought began in the Midwest. We did not realize how severe it was until later. It lasted through that year and until the end of 1936. It was later called the Dust Bowl – the years of the great dust storms in Kansas, Nebraska, and Oklahoma. However, it was just as bad in the other cornbelt states from South Dakota to Indiana. Our crops were a total failure for three consecutive years. The cinch bugs ate the com stalks until they looked like broom-handles in the field.

Our crops were a total failure for three consecutive years.

Editor's Note: It is impossible to truly comprehend the stress of the life of a farmer. You are at the complete mercy of the weather. I remember anytime I would spend the night with my grandparents, the first thing on the television every morning was the weather channel. Would it finally rain? Would it rain too much? Would it freeze too soon? Should I bale the hay today? Should I flood (or drain) the rice fields tomorrow?

Dad had kept up correspondence with a friend and rice farmer, Raleigh Robertson, in Katy, Texas. He was a man Dad had known when he was there in 1910. Dad wrote to him and asked him if he could find a rice farm for rent. He got a reply, and the answer was yes.

This was the opportunity Dad had wanted all those years to go back to Texas and farm rice. We were already in school in September 1936, but we all piled in the old 1928 Chrysler and headed for Texas to look the situation over and see if we wanted to make the move.

The decision was to move to Texas, so we went home, held an auction sale, and sold everything except one coop of chickens and some furniture. The farm itself was retained to fall back on if things did not go as planned.

Dad said if we got robbed, they would not think to look on a little boy.

Dad purchased his first tractor – a 10-20 International – with part of the sale money, and we loaded the tractor, chickens, and furniture in a box car and shipped it to Katy. I carried the balance of the money – about five hundred dollars – in the chest pocket of my overalls. Dad said if we got robbed, they would not think to look on a little boy. Then, we started back to Texas in the year 1936.

Editor's Note: The current distance by freeway between Grand River, Iowa and Katy, Texas is 888 miles (1429 kilometers.) Today, it would be about a 14 hour drive.

GENE ANDREW

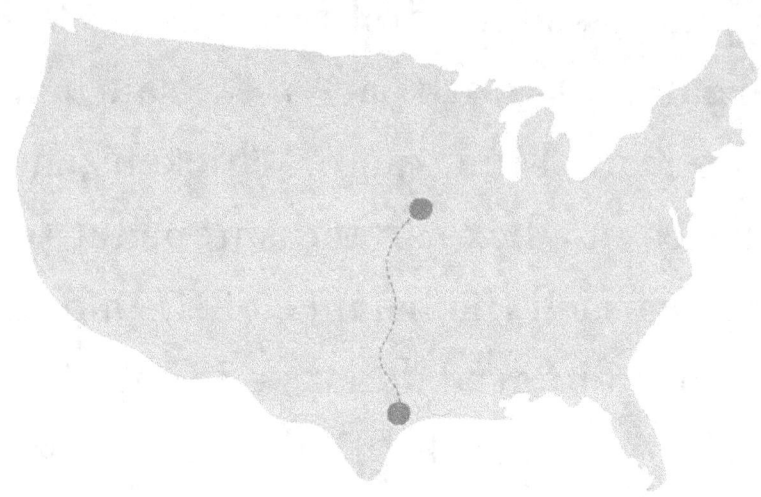

The Drive from Iowa to Texas

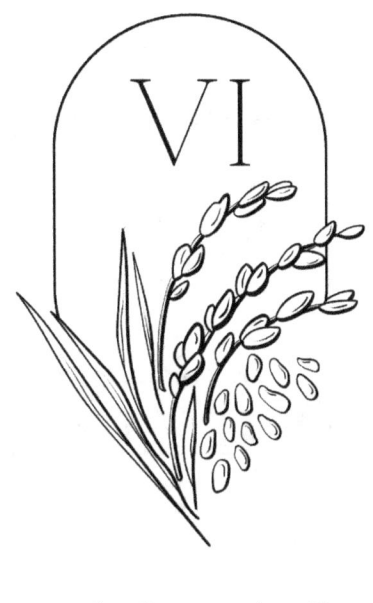

TEXAS

We arrived in Katy, Texas on my tenth birthday, November 16, 1936. The rice farm we had rented was known as the Weller place – a blackland farm south of Katy on Pin Oak Road, a mile west of Buffalo Bayou. (There is about

2000 acres of blackland south of Katy and about 500 acres at the southeast side of Cinco Ranch, about where Fry Road connects to Mason Road.) It had a small irrigation well and a square two-story house to live in.

Editor's Note: Blacklands are areas where the soil is black with heavy clay content.

Irrigating a Rice Field

Moving to Texas and to Katy changed my life forever and opened up a whole new world for me. The school was so big that I was sort of intimidated and pretty shy at first. My dad was so busy, he just dropped Harold and me off at the front door the first day, and we had to enroll ourselves. I was ten years old and in the seventh grade. The superintendent and principal took one look at me and said "no way". They put me in the fifth grade and Harold in the seventh!

My dad was so busy, he just dropped Harold and me off at the front door the first day, and we had to enroll ourselves.

I did poorly at first; I ranked 23rd out of a class of 24. By the fourth 6 weeks, I had moved up to #2 and the next one to #1. But a certain girl who had been at the top all year began crying, so Mr. Buening, my first teacher, changed her report card to #1 and mine to #2. This had to be my introduction to politics, however small.

After we got move in and settled down, Dad began plowing for the rice crop. When I got home from school, I would drive the tractor, plowing while Dad did other things. This was my introduction to real farming, not just doing chores. Right after, Dad saw the need for a more powerful tractor, so he bought a new Model "D" John Deere from Albert Thompson, Sr. who was a salesman for Harris County Implements Company. This was the best available tractor for rice farming of the '30s. The dealer in Eagle Lake sold over

500 of this model in one year. I will never know how Dad managed enough credit to buy the John Deere. We also used this tractor to pump water from the irrigation well.

The winter of '36 was a terribly wet and cold one. Pin Oak, being a black dirt road got sticky with deep ruts, and we could barely get to town. About that time, we started catching a school bus a quarter mile west; it was a long Packard automobile with three sets of bench seats, and held about nine. It had a heater and was real cozy riding to school in class. Later, they changed to a Model A Ford bus.

We only had 3 in 1 shoes (school, work, and Sunday.)

The only eventful thing that happened at school that year was when Harold and I got skunk odor on us, and since we only had 3 in 1 shoes (school, work, and Sunday), we went to school smelling like skunk. Harold got the nick-name Skunk and I was Skunk's Brother.

I thought this huge new school with one class per room was really neat. The boys played baseball and basketball and the girls played volleyball. I had never heard of these before and I was becoming quite adept at sports before the year was over.

At recess, we played washers and marbles, and everyone wanted to be my partner at washers because I could toss two in the hole every time. When I got through seventh grade, I had two gallons of marbles and lots of boys had none.

The next year, 1937, Dad, with all his energy, saw the need to farm a much larger amount of rice, and also get away from that blackland which is much harder to farm than sandy land.

Author's Note: Most of the land around Katy is part of the Katy Fine Sandy Loam, a designation in soil maps and geography books. It runs from about Tarkington, in East Texas, to Victoria. Rice has been grown in Katy since the 1890s because of the level terrain, easy availability of large amounts of irrigation water, and because there is a clay "land pan" about a foot down. When the topsoil is soaked with water, it will not penetrate the clay, and the only water lost is by evaporation. Katy is one of four premier rice areas in the U.S., the others being the Mississippi Delta; Stuttgart, Arkansas; and the Sacramento Valley of California.

So, we moved four and a half miles east of Katy to Mason Road, and Dad would farm 400 acres on Cinco Ranch. He was to farm on halves with a man named Rufus Hoffpauer who had rented a portion of Cinco for rice.

Mr. Hoffpauer was from Galveston and owned a rice mill there. His cattle foreman was Mr. Scallon. He and his wife lived in a large two-story house at Mason Road. (This house now sits at Porter Road and Franz.) Our house was about a half mile east of Mason Road and on the banks of Mason Creek. We were beside Highway 90 and MKT railroad and across the road north was Joe Moore's dairy.

Harold and I walked with Helen Moore to catch the school bus at Mason Road. While waiting for the school bus, to entertain ourselves, Harold and I would throw rocks from the railroad tracks. There were many telegraph and electric poles, and there wasn't a glass insulator left intact as far as we could throw. Also at that time, we laid a 1x4 board on the track to see what the train would do to it. We forgot about Mr. Bruton and his little crew of men. They rode a hand car and would keep the tracks in repair and also change cross ties. While we were at school, they came along and the board nearly overturned them. Mr. Bruton was waiting for us that evening when we got off the school bus and he was so mad, we thought we were going to jail.

The MKT had a siding about a mile long in front of our house; the freight trains would stop and take the siding to let the passenger trains pass. Every morning, there would be a freight train on the siding. In those days, they left the box car doors open and many hobos rode in them, so Harold would get in a box car and ride to school. I was always thinking it would go through Katy without stopping, but it never did; it would have been a good joke on him.

Editor's Note: Katy, Texas got the name "Katy" because of the Kansas to Texas railroad. MKT was the actual name of the railroad: Missouri, Kansas, and Texas. But it was more commonly called "The KT" or "The Katy."

House on Franz 1956 (Still Standing as of 2023)

SHORT TALES AND ADVENTURES

T his chapter is about the things we did while we were starting to grow up. When I say we, I am usually talking about Harold and myself. We were very close together in age and were as close as two brothers can be without being twins. During this time, our lives closely paralleled since we did everything together and were almost constant companions – in mischief and otherwise. Our memoirs should be exactly the same, but of course, they are not, as people will remember the same event differently.

I will also try to relate a few of the true stories handed down by my father. These events are in no particular chronological order as this would be impossible after so many years.

The Sheriff

Uncle Jesse was sheriff of Decatur County. He died under mysterious circumstances.

Uncle Jesse, the oldest of the family, was sheriff of Decatur County and appointed Dad as deputy sheriff for a while. While holding the office of sheriff, he died under mysterious circumstances. There were rumors that he was poisoned but this was never confirmed. However, the thing that made it such a mystery was that he was only thirty-six years old and in perfect health. Uncle Jesse died before I was born.

The Mill

Uncle Allen owned and operated the feed mill in Grand River. His main business was custom grinding, mixing grains for cattle and hogs, and grinding corn meal. We liked to go along to Uncle Allen's mill to get the corn ground. He had a large one-cylinder engine that powered the mill, and it was fun to watch him start because it went chug, chug, chug, and gradually picked up speed. There was a complicated system of pulleys, belts, and shafts all over the place to run the various machinery.

Always around the mill helping was his youngest son, Giffin. He was a big, good-natured guy who always took time to talk to us. Years later, we visited Giffin and his wife, Pauline, when they came to Port Aransas to fish and spend the winter.

One day, Dad let me go along to the mill in Grand River. He took several sacks of corn on a two-wheel trailer to be ground. When we started home, Dad gave me something to do to keep me occupied and out of his hair. He told me to get in the back seat, look out the rear window, and watch the trailer in case it came loose. There was very little chance this would really happen, but it did. The funny thing was, I saw it come loose but didn't say a word. When we arrived home, he saw the trailer came unhitched and I said, "You said to watch it. You didn't tell me to say anything if it came loose." Then we went back towards Grand River and found it a mile or so out of town with the tongue stuck in the hard ground.

We also liked to visit Uncle Allen at home because he kept a regular zoo of small animals in his backyard. He had squirrels, coons, foxes, etc. Aunt Lou had a house full of canaries which I think were sort of a hobby, but she also sold some.

Hogs and Strawberries

Uncle Lewis was the tallest and biggest of the Andrew brothers, standing about six foot three. He and Dad looked exactly alike, enough to have been twins, but Uncle Lewis was quite a bit heavier than Dad and was ten years older.

He was a farmer and raised registered hogs. Many of his hogs would reach waist high and weigh four hundred pounds or more. These were his breeding stock. He also was quite an expert in the culture of strawberries. The last time I saw him, in 1964, he gave us about a gallon of his very sweet strawberries to take along for the road.

His wife, Aunt Ollie, was a large woman, and they were a very jolly couple. She could fairly pound the piano and read no music while playing.

They had one son, Johnny, who went to school with Doris in Grand River.

The Other Farm

Uncle Harry was my favorite uncle. He lived the nearest, and I believe was closer to Dad than any of the brothers. He lived about a mile and a half northeast diagonally from us, as the crow flies, but if you went on the road, it was a half a mile north and a mile east. We were allowed to visit Uncle Harry's often by ourselves and would walk or ride the pony across the field. Uncle Harry's farm was one hundred sixty acres also. My grandfather had divided his property equally among the children. We went to visit our cousins: Ray, Ralph, Vernon, Dorothy, and Donald. Donald was about our age, so he's the one we played with around the farm.

This is probably one of my best recollections of Iowa.

Uncle Harry had two large barns that held lots of hay. The hay was put in these barns by a mechanical hay fork that picked up the hay on the ground outside the barn, traveled vertically up to the top, and then horizontally inside the roof peak. This was powered by a horse walking in a straight line. There were many pigeons in these barns, and I loved to find their nests and look at the squabs. This is apparently where I developed my interest in pigeons.

There were many other things to do there, and this is probably one of my best recollections of Iowa. Donald and Dorothy often came to our house via the same route. Dorothy and Doris were pretty good pals. Donald helped us get into mischief quite a few times. Ralph and Vernon were a little older than Doris, but they loved to aggravate her to no end.

One hot summer day Harold and I started out to visit Donald. The folks had gone to

Leon. We had some cattails that we had soaked in kerosene and were carrying them like torches. We were about one-half mile from home and looked back, and the whole pasture was on fire behind us. We ran back to the house and got some wet gunny sacks but weren't *about* to be able to put the fire out. It burned itself off when it got to the road. A few days later, we overheard Dad say to Mama, "I just can't figure out how that pasture got burned off."

Uncle Harry lived to the age of one hundred one.

Lead Poisoning

I suppose we visited the least with Uncle Arthur. The reason we did not visit more was that he operated an automobile repair shop in Leon for a while. He was the one nearest in age to Dad.

Dad told many stories that included Arthur, but the one I remember best was the one about when he was cleaning a gun. He fastened a heavy wire cleaning rod in a vise and was sliding the gun barrel back and forth on the wire instead of pushing the wire and cleaning material into the barrel as he should have been doing. Once, he pulled the gun back too far and missed the barrel, pushing the wire nearly through his hand.

While Uncle Arthur ran his garage many years, and due to the nature of the mechanic business, his hands were greasy over and over again. He would always wash his hands in leaded gasoline which was about the only kind available. He began to have severe headaches almost constantly and finally had to close the business. The medicine he took for the headaches was Bromo-Seltzer. He had bushels of those blue bottles behind the house. It was thought that he died from lead poisoning, but it was never confirmed. He died at age fifty-nine.

World War I

Uncle Paul was the only one of the brothers who served in the armed forces during World War I. He was in the Navy aboard the battleship Arkansas and was powder man on one of the sixteen-inch guns. It was said that Uncle Paul, being a large, powerful man, could slam the powder into the cannon better than anyone else in the Navy.

Uncle Paul's only son, Earl, went to California shortly after getting out of school. He married there and between the years 1960 and 1969 he, and his wife had eight children. Among these were a set of twin boys and a set of twin girls. Earl died at age thirty-eight. Uncle Paul lived to the age of seventy-four.

Editor's Note: The brothers who did not serve in the War were likely exempted because of their agricultural roles, or, in the case of Arthur, because of poor health.

Clyde Barrow

Probably the most noteworthy event of my early life was when Clyde Barrow and Bonnie Parker came to our house. Clyde stayed and worked for Dad for about three weeks. They were some of the most infamous outlaws of the 1930s, and this must have taken place in about 1933.

Harold and I had been to the corn crib, playing. We were about seven and nine years old. The corn crib was about two hundred yards east of the house, and there was a slough between the house and the corn crib. When we started home and reached this little slough about half-way home, we met two men walking towards us. One of them asked where Dad was, and we told them he was over in the cornfield cultivating and pointed to him. They proceeded on to the cornfield, about three fourths of a mile away, and asked Dad for a job. It just happened he needed some extra help about this time, so he hired them.

However, one of these "men" turned out to be a woman in disguise. Even at this age, I could recognize that this was a woman because of the slight build, her small hands and feet, and the facial features of a woman. She had a man's haircut and wore men's overalls. She left that same day and never came back. We never saw her again after that. Of course, this was Bonnie Parker.

The man was Clyde Barrow. He stayed three weeks and was a very good worker, doing everything Dad asked of him without complaining. He was working for his room and board and a small wage.

> One time, he was mowing hay in
> the field and ran over a quail's
> nest. He gathered the eggs into
> his hat and brought them home.
> He sat them under an old hen and
> hatched them out.

One time, he was mowing hay in the field and ran over a quail's nest. He gathered the eggs into his hat and brought them home. He sat them under an old hen and hatched them out.

Another time, he was milking and Dad said to him, "You look kind of like a jail bird." This was because a convict's hair was cut in a distinct way so that people could recognize them by this if they escaped from prison. Clyde jumped up from the cow and left the milk bucket under her, very excited, and demanded to know why Dad had said this. Dad told him he was just making conversation, and Clyde was finally convinced Dad meant nothing by it.

Clyde was always ready to make homemade ice-cream, which we did a couple of times while he was staying with us. He would crush the ice, turn the handle, and eat his share.

He wrote several letters to a woman in Arkansas, later believed to be Bonnie Parker. He would always ask my sister, Doris, to address the letters, something we thought very odd at the time, because if he could write letters, he could have addressed them. We later realized he was keeping his handwriting out of the mail so the authorities would not see it.

Finally, he left without telling anyone he was leaving, and forged checks on Uncle Paul and some other people in the neighborhood. He tried to forge one on Dad in Kellerton, but the merchant knew Dad's handwriting and refused to cash it. Clyde left the bank without arguing.

Editor's Note: According to Gene, it was not uncommon to have people pass through from time to time, work for an indeterminate amount of time, then leave unannounced.

Then, about six months later, one morning Mama opened the paper, *The Des Moines*

Tribune, and there was this man's picture on the front page, and also Bonnie Parker. It was an article telling about the ambush and death of Clyde Barrow and Bonnie Parker by officers in Louisiana. Mama immediately recognized their pictures and said, "This is the man who worked for us."

Later she related this story on a radio talk show in Houston, and the host kept her on the air a half an hour telling about our experience.

Going to Reunions

Some of the most popular events in the area were the reunions held in each town in the fall of the year. We usually attended the ones in Grand River, Decatur, Kellerton, and Davis City. They were actually more like a county fair. There would be hundreds of people in attendance. There would be a huge carnival with all the rides and ball throwing stands. Then there were the side shows of small circuses, "freak" shows, and the peep shows for the young guys. (I believe these were strip shows, but children could not go in.) Then there was the balloon ascent and parachute jump each day and airplane rides.

I was fascinated by the airplanes. One reunion had three planes, each a different color, and I watched them land and take off again and again. Sometimes they would fly directly overhead at about five hundred feet altitude.

These reunions were much better and more entertaining than the county fairs of today. My parents would see many people they knew and hadn't seen for a long time.

Dad let me ride the little cars once, and I ran over a man standing beside the little racecourse. These were little cars of different colors, each with a small engine mounted on the rear, much like a go-cart. The oblong raceway was marked off with ropes on stakes on each side, and I was supposed to go around a few laps between the ropes. Right off, I lost control and also stepped on the gas too much. I shot out from under the ropes and off the track, and there was this man standing there. He made a jump to get out of my way, but it was too late.

Homemade Ice-Cream

Once in a while, we would decide to make homemade ice-cream. This was not the simple thing it is today. First, we had to bring in one of the cows in order to have fresh milk. We had never heard of homogenized. Then, while Doris and Mama stirred up the custard, Dad, Harold, and I would go to Grand River to the ice house for a fifty pound block of ice.

The icehouse was a huge old two-story building and was full of ice cut from the river with big crosscut saws, then packed in sawdust in the icehouse. It melted very little this way in the summer.

When we got home, Dad would crush the ice with his double-bitted ax, hitting it inside a gunny sack. We used salt that had been bought for the cattle.

Editor's Note: Salt was given to cattle as an electrolyte supplement to replenish their salt lost by sweating.

The finished product was delicious, and we never ate cake or anything with it- just ice cream. We usually made banana, pineapple, or vanilla.

Firecrackers

The Fourth of July meant it was time to shoot firecrackers. I was surprised that Dad would buy them for us, let alone shoot them, but we could have a package or two of the medium-sized ones. Each one had a particular purpose such as blowing up tin cans or Prince Albert tobacco cans. We had very little supervision at times.

Editor's Note: I also recall Gene telling me about putting some in a mailbox and blowing it sky-high. They definitely got in trouble for that one.

Silent Movies

In Decatur, a small town about halfway to Leon, they showed silent movies on Saturday. We would go a few times a year. It was all outdoors. There was a big white screen up front with many rows of flat, wooden benches for the viewers to sit on. Then at the rear was a projection booth – a little square building twelve feet high mounted on four legs with a little ladder going up to the entrance.

The movies were mostly "Cowboys and Indians." It would show a sequence, and their mouths would move. Then it would show in print what they were saying. As a reel was finished, there would be about a five-minute intermission while the projector operator put on a new one. I usually fell asleep before it was over.

I saw my first and only talkie movie at the Leon Theatre just before we moved to Texas. It was about the Dionne Quintuplets and Dr. Defoe who delivered them. I enjoyed it very much and wondered how movies could talk.

Decatur also had a drug store that Harold and I liked to look around in. It had about everything there was on the market in the way of medicine, ointment, salves, and old-time remedies. They also had things like knives, watches, etc. This is where we could get a double dip ice-cream cone with a third dip on top. These trips to Decatur were the highlight of our meager existence and eagerly looked forward to.

Illnesses

As related before, we were very seldom sick, considering the wintertime conditions. We did have about all the childhood diseases, but I never had an inoculation of any kind. We had the mumps, measles, chicken pox, and I don't know what else. When we had the measles, we got over it in the usual time. However, it left me with some kind of eye problem. I could not see, and my parents were sure I was blind. I could see a little, but the light hurt my eyes so bad I could not open them.

The doctor said the measles had settled in my eyes.

It was decided I would be taken to an eye specialist in Creston, Iowa who was the best in the State. Whatever he did, and the medication he gave me, must have helped because I could read the highway signs on the way home. The doctor said the measles had settled in my eyes.

Another time, Harold and I broke out in a severe rash. We had to put a sulfur mixture on our entire body, then put on long handle underwear to keep from rubbing it off. We had to stay home from school two weeks, but we were so miserable that it was no fun, and we might as well have been at school. We never knew for sure, but blamed it on buckwheat pancakes. Dad had grown a field of buckwheat that year and had some of it ground into buckwheat pancake flour. For a few months, we had buckwheat pancakes every morning for breakfast.

Learning to Swim

All the neighbors from all around brought their teams.

Just north of the orchard, about two hundred yards, was a slough which Dad made into a stock pond. All the neighbors from all around brought their teams and Fresnos – a 1/3 yard dirt scraper that was pulled behind horses with the operator walking to manipulate it and make it fill with dirt. After quite a few days, they had dug out a very large hole that would hold water about ten feet deep in the center, and the dirt dug out was used to build the dam. Right in the center, he constructed a little platform about six feet square on tall, wooden legs that would stick out of the water. Then when it rained, the pond filled with water to the spillway.

In the summer, Harold and I would go to this pond, take off our clothes, and go swimming. We learned to swim, and Harold would swim out to the platform in the middle of the pond, but I was not that brave, so I stayed around the edge. Mama never knew that we were going swimming.

Learning to Smoke

When we were about seven and nine, we decided that we were old enough to smoke.

When we were about seven and nine, we decided that we were old enough to smoke. Dad kept his Prince Albert tobacco and cigarette papers with him all the time, so we could not use real tobacco. We tried about everything that would burn. First, we tried ground leaves, sawdust, etc. but none of those were very good. The best thing we ground was dried willow roots and dried grape vines cut about four or five inches long. Both of these would burn and stay lit, and they were porous and would draw. We smoked for a while, but soon gave it up as it wasn't worth the trouble.

We could make an excellent corn cob pipe in just a few minutes. Just cut a corn cob off the right length, clean out the center with a pocket knife, bore a hole in the side, then cut a round, dried weed and push out the center core with a wire, and you had the stem.

When I was a little older, Mama told me, "You mustn't smoke as it is a bad, nasty habit and bad for your health." I took her at her word and never smoked again. Dad chewed tobacco and smoked his Prince Albert as long as I can remember – until he was very old – so it is a wonder that we didn't also.

The Slingshot

We became very good at making slingshots. You just cut a good, solid tree fork about the right size, made notches at the ends of the forks, then cut the rubber strips from an old inner tube. The trick was to cut the strips of rubber the same length and tie them on equally. Then you cut a leather rock holder from an old shoe tongue and fasten it in the center of the rubber strips.

Some of these slingshots were more accurate than others. The best one that we ever made was confiscated very quickly. Harold took a shot at a hen – it happened to be one of Mama's best ones – and just as he shot, the hen turned her head to the side with the rock hitting her in the head, killing her. She went flopping all over the place, and it all happened just as Mama looked out. She grabbed our slingshot and threw it in the fire (the kitchen wood stove) and burned it up.

The Fire

Speaking of that old kitchen stove, one time I took a long, dry stick and put it through one of the vents in to where the coals were on fire and set the stick afire. I went to a long drape-like curtain near the stove and set it afire. I can't think of any good reason for doing this – maybe I was a budding arsonist. Anyway, the curtain flashed up in an instant and was burning all the way up to the ceiling. Mama saw the fire and yelled at Dad who was out by the driveway visiting with a neighbor who had just driven up. Dad made about ten long strides to the house (I never saw him move that fast before), grabbed the water bucket which just happened to be full of water, and doused the fire. This was one thing that I did on my own, I didn't have any help from my brother.

The Hobo

One year, my folks decided to raise chickens. We bought a couple thousand chicks and converted the hog house to a brooder house. As soon as the chicks were three or four weeks old, they were turned loose out into a pasture in one big flock. Immediately, they attracted chicken hawks, so Dad had to almost stand guard over them as the hawks were raiding them pretty badly. He didn't have time to watch them all the time, so he made a scarecrow out of straw, put a shirt and overalls on it, and an old straw hat and propped it up on an old, overturned stock tank. He placed it out near the chickens, and it really looked like a person sitting there.

The first morning, we overheard Dad (he said it where we would be sure and hear him) say that an old hobo came by early that morning and he hired him to guard the chickens in return for his meals. He told Harold to go out and tell him breakfast was ready but also said the old man was very hard of hearing. Harold went out towards him and hollered "breakfast's ready." Dad kept telling him to get closer and closer, and finally Harold was so near he could see his face. He was completely put out and ambled back to the house real slow. Mama fussed at Dad for playing such a dirty trick on a little boy.

The Shotgun

Dad kept an old single-shot shotgun out in the barn to shoot chicken hawks. He kept it loaded and thought he had it hidden good from us. Harold and I would make regular visits to this old gun after school to play with it. It was the kind of gun that breaks open at a pivot point between the barrel and the stock. Harold could break it open and remove the shell, then we would click the trigger several times. One day, I went by myself to play with the gun, but I couldn't break it open to unload it. So, I just pointed it upwards, put the butt against my stomach, and pulled the trigger anyway. I blasted a big hole through the roof of the barn and tore a hole in my new overalls from the recoil. I had a real good skinned place on my stomach.

All the rest of the family was in the house at the time. They were all out to the barn in a flash, but Mama and Dad were so glad to see me still alive that I didn't even get a whipping.

Other Short Stories

I don't remember this one, but Doris told me about it. One year, there was a deep snow drift a few feet north of the house. Harold and I dug a tunnel several feet back into the snow and would play in it as it would be warm in there. One day, the tunnel collapsed on us, but, somehow, we managed to escape. We went in and told

Mama what had happened, and she was very upset with us.

When Wayne was about three or four years old, he was always wanting to ride Scotty. He managed to get on her bareback and with no bridle. She took off out of the bam at full speed, and when she rounded the fence comer at the northeast comer of the orchard, he fell off and went rolling over and over. She came back and stood beside him so he could get back on.

I was fascinated by the trains in Grand River and loved to see them go past. Whenever I could go along to town with

Dad, usually to Uncle Allen's mill, we might be lucky enough to have a train go through. Although passenger trains were the principal means of travel, I had never been on one.

On one occasion, Mama took me along to visit Mrs. Smith, who lived up the road a half a mile. Mrs. Smith said, "I would think you wouldn't want your boys to go to school at East Elk because there are some pretty unsavory characters going there, and your boys will learn to swear." I heard what she said, and I told her, "We don't have to learn it; we already know how." I don't know why I said this because it wasn't true.

One day, we got a phone call from Aunt Hazel telling about our cousin, Myron, falling into the water well. Their two boys, Myron and Don, were about like Harold and me in age and played together all the time. This particular day, they were playing near the well, which was a well dug by hand about four feet in diameter lined with stone and about twenty feet deep. On top was a hand pump (this was modem times and the oaken bucket was out) mounted on a wooden cover. The boards had gotten rotten, and Myron fell through to the bottom. Myron clung to the side while Don went to the house and calmly said, "Rat fell in the well." They didn't know who he meant as he never called him by that name before. Uncle

Clyde ran to the well with a piece of rope and pulled him out.

Mama and Dad were talking one day about a man named Mr. Tongate who owned property nearby. Dad was saying Mr. Tongate had cancer and would not live very long, then they went on discussing other people they knew who also had cancer. I often think of this when I hear people blaming pesticides and fertilizer for causing cancer.

Editor's Note: I would have loved to pick Gene's brain about this. Gene himself died of mesothelioma. Other than his brief smoking attempts as a child, he never smoked. He likely developed lung cancer from being in contact with fertilizers and pesticides as a farmer or from serving in the Navy where he was exposed to asbestos.

Christmas

> **The only day of the year we recognized as a holiday was Christmas. We never celebrated birthdays or any other holidays.**

The only day of the year we recognized as a holiday was Christmas. We never celebrated birthdays or any other holidays as they were just another day of the year. There would be a Christmas program at school or at Elk Chapel where each child would get a bag of fruit and nuts.

Elk Chapel

At home, we got one present each on Christmas morning. One year, I got a toy Caterpillar with rubber tracks, and you could wind it up, and it would go. I was very proud of this toy but the very first night, I left it wound up tight and the cold weather caused the spring to break. I only got to play with it one day.

I recognized Christmas as a happy time, and I still look forward to Christmas just like a child. At this time, all I knew about Christ and God was what my teacher taught at school. I didn't relate it to being Christ's birthday.

In the winter, we spent many hours sliding down hills on our sled. This was great fun, but it was hard work pulling it back to the top of the hill. I had not learned to ice skate and Harold was just learning when we moved to Texas.

Dad's Stories

A great many of the events Dad told us about were the things that happened in World War I.

Although Dad did not go into the service, many of his friends did, and they related many things about the war. The thing that impressed me the most about the war was the cold and hardship that the soldiers endured in France, going "over the top" of the trenches to advance, and the use of mustard gas. During battle, anyone who got the faintest smell of gas yelled out "gas" and everyone in hearing distance put on their gas masks. The first person to smell it usually died within minutes, but if the gas was diluted somewhat by air, the lungs were permanently damaged for life.

Dad's favorite stories were about his own physical prowess. I think the young men of his time actually fought and wrestled purely for the pleasure of it and for recreation. The young guys would come from all over and congregate in Grand

River on Saturday night, and it nearly always ended up in a fight. There was no drinking which would be just about unbelievable nowadays.

There was tremendous competition beginning with horseshoes, of which Uncle Lewis was champ. He could throw a double ringer every time. He competed at the state fair in Des Moines for the state championship.

Other things they did were weightlifting, not like today, but actual pounds to see who could lift the most, wrestling, and "Indian pull-ups." "Indian pull-ups" was a very physical contest where two men sat down facing each other with their feet together. They would take hold with both hands of a broom-handle above their feet and then see

who could pull the other one up. Dad was
the champ of this and could pull up men
much heavier than he. He lost one time and
that was to Uncle Lewis.

Wrestling was collegiate style, not the
fake TV style. I think it was pretty
well conceded that Dad was number
one in pure strength, and that was
a great accomplishment considering the
competition. Everyone was in good shape
from hard, manual labor on the farms.

One time, we had Aunt Helen's family over
for Sunday dinner. This was Mama's sister
and brother-in-law who had a large family
of boys who were small in stature but were
huge eaters. They could each eat as much

as two or three people. Dad was very fond of squirrel, and we had a big platter of it that day. Just as dinner was beginning, a car drove up, and Dad had to leave the table for a little while. When he came back, there was only one piece of squirrel left on the platter. Dad reached to get it with his fork, and Fritz Rhodes speared it with his fork right out from under Dad's fork. To say that Dad was furious is the understatement of the year.

In the winter, the streams would freeze over solid with ice. Dad and his brothers and friends would get on the Grand River, which was nearby, and ice skate many miles up or down the river.

They did a lot of hunting and trapping, and the small game was easy to trail in the snow. They did a lot of fox and rabbit hunting. They trapped and sold furs for extra money.

A man in the neighborhood, Orrin Stanley, was a professional beekeeper, i.e. he sold honey for a living. If Dad had some spare time, he helped Orrin tend to beehives all over the county, and we also had about two dozen hives on our own place. Dad gained a thorough knowledge of the science of beekeeping and passed it along to us.

One form of recreation was house dancing. There would be a group of friends who would meet at a particular house for a dance, and it would go 'til after midnight. Dad enjoyed playing the banjo in a small band that played for many of these dances.

Whomever was the successful bidder on the box lunches would get to eat it with the girl who made it.

The only fund-raising activity was the box supper that was usually held at the school or Elk Chapel. The ladies and girls of the community would make a box lunch or pies and cakes. These would be auctioned off to the highest bidder, and whomever was the successful bidder on the box lunches would get to eat it with the girl who made it. There was always a lot of good-natured spying and finagling going on amongst the men to find out which box belonged to whom.

Dad disliked the concept of insurance and loathed insurance salesmen even more. He never had an insurance policy of any kind. There was a certain insurance salesman that began calling on him, trying to sell life insurance. He was very persistent and would show up every few days. Finally, Dad told him in no uncertain terms to not come back anymore.

Then, one day, Dad was up on a ladder
painting the barn when this salesman
drove up. Dad's blood pressure shot up, and
he came down off the ladder. Dad turned
the salesman around and kicked his rear
every step of the way back to his car.

Many of Dad's stories involved his Uncle
Luther who came to Iowa from Palmyra,
Illinois. He married a sister of Grandma
Evaline Akers, so Dad had double cousins,
too. Also, he told many stories about
Spencer Collings Akers and his brother,
Benjamin.

I never knew much about mother's side of
the family. She was not a good storyteller
and most of her family lived a much greater
distance away. Dad traced her family tree
far back in his little black book which
he started in 1923. They were some of
the early pioneers of Southern Iowa, and
some of the family names were Brown,
McDowell, Wilcott, and Elmore.

THE WAR

During "The War" (WWII) it was necessary to ration such things as sugar, coffee, meat, and other groceries. Then there were rubber boots, shoes, etc., and the bigger things were tractors, farm implements, cars, tires, batteries, gasoline, diesel, etc.

The Tractor

Dad managed to get a permit for a tractor, so he found a nearly new Farmall up near Livingston, Texas at Goodrich. He sent Harold and me after it (about 100 miles) in an old International Buick we used around the farm. We found the right place and paid the money and managed to get the tractor loaded and tied down. (We were about thirteen and fifteen.) We started home on Highway 59 and had a double blow-out a few miles south of Goodrich. It was getting dark, but we found a home where they let us use the telephone. So, we called Dad who said he would come right on over with two tires.

We waited all night, and the humidity and mosquitoes were terrible. Dad had brought the tires to Humble and then up 59 but was not going far enough. He stopped at an all-night cafe in Humble where a bus was stopping. He asked the people that got off if they had seen a truck with a tractor on it. Some of them said yes, and that it was a long ways back. Then Dad knew to go further north. We got home just at sun-up.

Gene Andrew and High School Classmates

High School

When I was in high school, I became very interested in sports. Besides the high school team, Katy had two men's basketball teams both of which were very, very good and seldom ever lost a game. They were nearly professional in ability. I would stay after school and hang around until the game started and was willing to walk home the four and a half miles to Mason Road just to get to see them play.

One night, after the game, I went to Alexander's café, where everyone hung out, to see if I could catch a ride. A man named Clarence Pennington said he would take me home, so he did, and I thanked him. The very next day he was killed on a tractor that turned over on him. (The names of the basketball teams were Katy Methodist and Miller Motor.)

Playing Trumpet (Second Row, Center) at High School Dance

1943 Commencement

OF

Waller High School

Baccalaureate Sermon

Waller High School Auditorium
Waller, Texas

MAY 16, 1943   11:00 A. M.

*Waller High School Commencement –
1943*

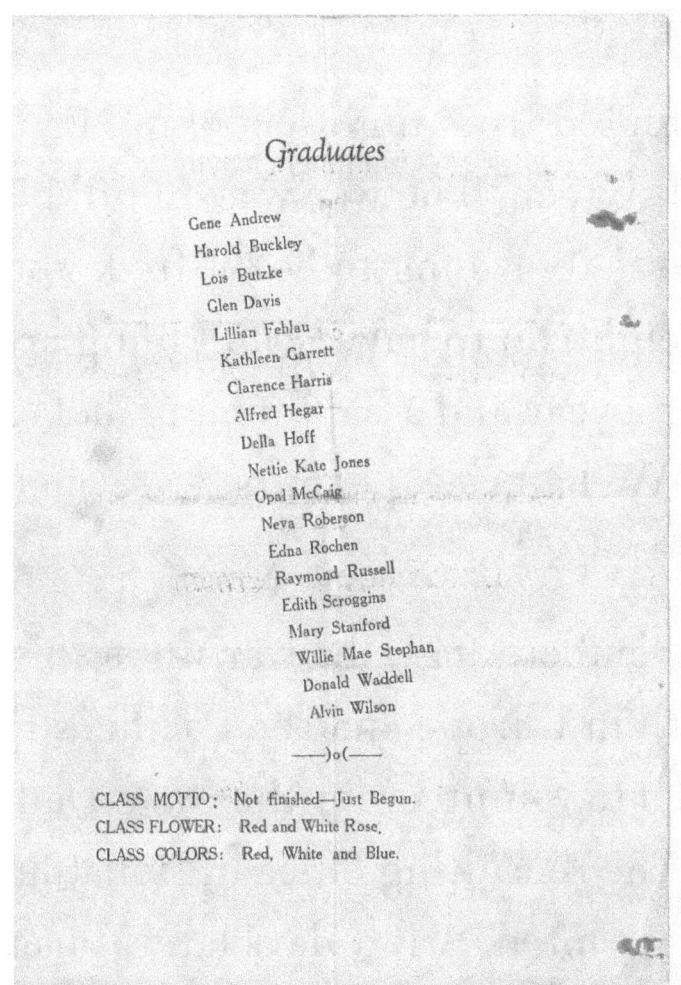

Graduates

Gene Andrew
Harold Buckley
Lois Butzke
Glen Davis
Lillian Fehlau
Kathleen Garrett
Clarence Harris
Alfred Hegar
Della Hoff
Nettie Kate Jones
Opal McCaig
Neva Roberson
Edna Rochen
Raymond Russell
Edith Scroggins
Mary Stanford
Willie Mae Stephan
Donald Waddell
Alvin Wilson

——)o(——

CLASS MOTTO: Not finished—Just Begun.
CLASS FLOWER: Red and White Rose.
CLASS COLORS: Red, White and Blue.

Waller High School Commencement – 1943

Draft

I graduated from high school in May 1943 at age sixteen. The war was on full blast, but I was too young to be drafted. You had to be eighteen to enlist. So, I hung around Waller a year and a half helping dad on the farm. We had rice, peanuts, watermelons, and cattle.

By November 1944, the war was still going on both in Europe and the Pacific, so I decided to volunteer for the Navy. Dad took me to the recruiting office in Houston, and I was accepted. A few days later, I had my physical exam. Then I had a week or so before I would leave for boot camp in San Diego. I was still seventeen.

Editor's Note: I do not remember what Gene specifically told me about his early enlistment. Either Gene and his father put a different date for his birthday (so he was of age at the time of enlistment), or the Navy went ahead and accepted him since he was about to turn eighteen.

Gene Andrew – in the Navy

During my final week at home, all my friends gave me a "going away" party at our house. One of the girls who came to the party was Ora Lee Knebel, and she brought along her little sister, Dorothy, about thirteen years old, and I thought to myself, "Why did she bring a little kid along?" Little did I know, at the time, she would be my future lovely wife.

Editor's Note: Gene and Dorothy eloped December 31, 1954.

Dorothy Knebel Andrew

Gene Andrew – Navy Picture 1944

Lyman I. Andrew & Sons

RICE PEANUTS CATTLE

Waller, Texas

work unloading and loading most of the day. Don't
expect to sell any. We grab out our knives and all
hands cut a while then work at something else. We counted
the rows as we left the field this evening and found that
we still have over half of them left the first time over.
We thought we were doing fine till then. But we are
down on the north end and they aren't near so good.
Today Doris called and said she wasn't going to
work tomorrow and wanted mamma to bring the
baby and come down. So Harold and her left a while
ago. Think Doris expects to keep the baby there for
a while anyway. We have had a Mrs Morgan working
here the past several days and she is a good hand
and appears to be very nice. She also left tonight.
Wanted to tell you that we have had pretty hard
going in the melons on account of soft ground. We
use the cat ahead of the trucks most of the time
and today pulled one of the looks out on a cable
so maybe that will give you some idea just how
soft the ground is. But a truck tries to mire down
pretty bad with 8 or 10 thousand lbs anyway.
I don't think I feel any too hot tonight. Am
as tired as can be to start with, then galded under
both arms till I can't let my arms hang down, then
when I built that wagon I used creosoted lumber
and got that on my arms and face and they

Page from Letter from Lyman Andrew
to Gene Andrew During the War

We were sworn in (about twenty-five of us)
in Houston, and in a few days, we left for
the San Diego Naval Training Center (boot

camp) on the Southern Pacific. In El Paso, we were joined by about fifty recruits from the Dallas area.

It took forever to get there because the train stopped so often. At Yuma, Arizona we had a thirty-minute stop, so everyone got off the train. When we started on, there were five or six that were left in Yuma. At San Diego, we boarded a bus to take us to the Naval Training Center.

When I got off the bus, I wondered to myself, "Who keeps this place so clean?" It was not long until I found the answer to that question. We would hear the bugle at 4:30 every morning. We scrubbed and cleaned the barracks and marched to breakfast to one of the four chow halls before daylight. The schedule for the day was posted by then, and we were kept busy all day going from one thing to another. If there was a thirty-minute opening, we either hit the grinder (large open asphalt area) for drills or went to the obstacle course.

> **They always called us "dumb Texans" and we called them "prune pickers" which made them very angry.**

Each company was made up of 164 men divided into two platoons. We had enough from Texas for one platoon, and the other platoons were California boys. They always called us "dumb Texans" and we called them "prune pickers" which made them very angry.

Gene Andrew – Center

Aircraft Carrier

Boot camp was an incredible experience, at least it was for me. The physical conditioning part was superb. You met all kinds of people: some were just great, and some were the opposite. Some were real comics with good personalities and fun to be around – those were the kind you wanted to go on liberty with. "Liberty" was a twelve-hour pass.

We had one guy who would not shower, and that is really a no-no because disease can start so easily where people are confined so closely together. One day, about six or eight guys caught him and stripped him to his shorts and skivvy shirt and dragged him to the showers. They took the kiwi brush and soap to him and literally scrubbed his underclothes off of him. He was pink as a baby when they got through, and he showered regularly after that.

Flight Deck with Fighter Planes

From nine to ten at night was reserved for letter writing and then "lights out." Some of the things we did were firefighting, ship nomenclature, boats under oars (in the harbor), rifle range, twenty-five mile hikes to Oceanside, drill on the grinder for hours, boxing, swimming tests (first, second, and third class), and lectures of all kinds. During a lecture you had better not go to sleep.

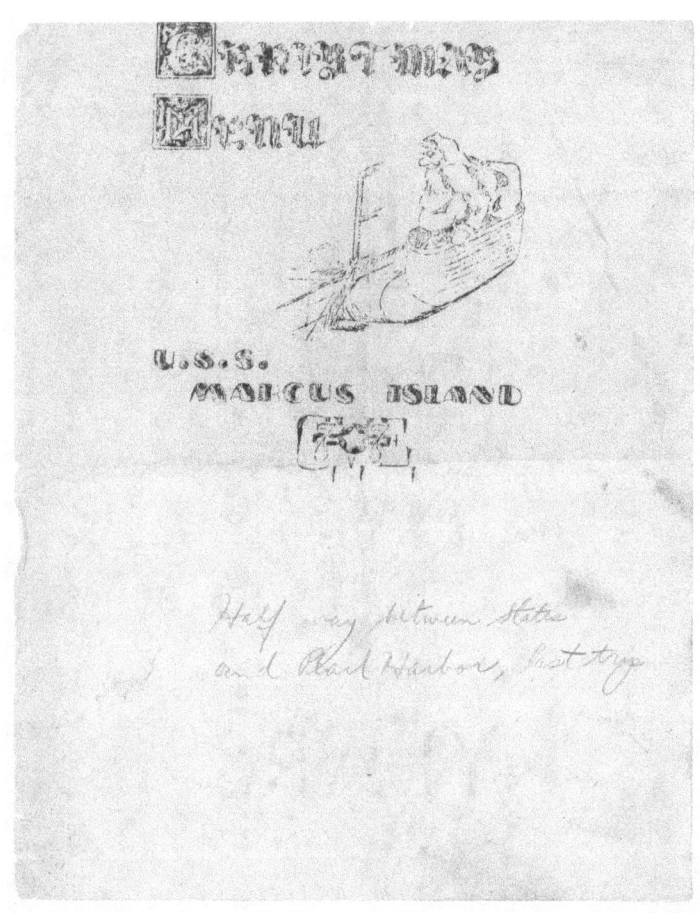

Christmas Menu, U.S.S. Marcus Island
"Halfway between States and Pearl
Harbor, last trip."

YEARS

About 1950

Waller Merchants Defeat Cleco, 14-4

WALLER, Aug 29. — (Sp)—
The Waller Merchants defeated
the Cleco Club of Houston 14-4
here Sunday. Gene Andrew
pitched six-hit ball for the win-
ners and had good support
afield. Royce Schultz featured
the Waller 14-hit attack with
four hits in five trips, including
two doubles, while driving in
five runs.

CLECO	Ab	H	Po	A	WALLER	Ab	H	Po	A
Cumby,1b	5	0	9	0	Armer,ss	5	2	0	3
Alley,2b	5	1	3	4	Smith,2b	4	1	2	2
Glaznec,cf	5	2	2	0	W.Schultz,1b	5	2	12	0
O'Neal,ss	4	1	3	5	D.Cubstd,3b	3	1	0	2
Bobb,3b	4	0	1	1	Grader,c	5	2	4	0
D.Rose,lf	4	1	2	0	R.Schultz,rf	5	4	3	1
Stevens,rf	4	0	0	0	Clark,lf	4	0	4	0
H.Rose,p	4	1	1	5	H.Cbstead,cf	5	2	1	0
Baker,c	4	0	5	0	Andrew,p	4	0	1	9
Totals	39	6	24	15	Totals	40	14	27	17

Cleco .... 101 000 110— 4
Waller .... 730 002 20x—14

Runs-Armer 2, Smith, W. Schultz 3, D.
Cubstead 3, Grader 2, R. Schultz, H. Cub-
stead, Andrew, Cumby, Alley 2, D. Rose.
E-Armer, Smith, D. Cubstead. R. Schultz,
Andrew, Glaznec, Bobb. RBI-R. Schultz 5,
Grader 4, W. Schultz, Clark, H. Cubstead 2,
Glaznec 2, O'Neal, H. Rose. 2B-R. Schultz
2, W. Schultz 2. Left-Waller 8, Cleco 10.
BB-Off Andrew 2, off Rose 4. SO-By An-
drew 4, by Rose 3. U-Cubstead and Daily.
T-2:30.

*Newspaper Clipping – Gene Baseball
Highlight*

1953: 27 Years Old

Article Highlighting Gene and Lyman Andrew – 1953

1956: 30 Years Old

In 1956 we had rice in Polk County by the Trinity River. We had cleared about 300 acres of hardwood timber, so the rice field was in the middle of a forest you might say. When we were harvesting, we had to take the rice out of the field over a wet, low place with the carts, and it was getting boggy, so we sent Lawrence and another helper into the woods with a chainsaw to cut several small logs to throw into the low place.

Pretty soon we heard Lawrence yelling for help. We thought something bad had happened. We ran into the woods and found them. They had killed a huge rattlesnake. It was about three inches in diameter but about midways on his body was a place much bigger and about a foot long. Harold just had to know what was in that snake, so he cut him open, and it was a full-grown squirrel!

About 1960

Newspaper Clipping from The Chronicle

1961: 35 Years Old

MR. AND MRS. GENE ANDREW
He's Outstanding Farmer of Year

Rice Grower Andrew Is Top Area Farmer

Gene Andrew, 35, rice and cattle farmer from Katy, is the outstanding farmer of the Houston area for 1961.

He was so named at the annual farmer awards dinner of the Houston Junior Chamber of Commerce Saturday night at Bill Williams Restaurant, 6515 Main.

Andrew was one of four nominees for this honor. The other three were A. L. Furnace of Cypress; Wesley Nelson of Katy and Aubrey Chudleigh of Hockley, all are rice and cattle farmers.

With borrowed money, Andrew began farming in 1947, after serving in the Navy during World War II. His assets included two used tractors and a used plow.

He obtained a loan from A. W. Umland, president of the Guaranty Bond State Bank of Waller.

Andrew first leased 350 acres from the Harold Longenbaugh Estate at Katy in 1947.

Today, he operates on about 5000 acres. About 3800 acres are in Polk County near Onalaska for his beef cattle operation.

The other 1200 acres he leases are about six miles northwest of Katy.

The Polk County land is held in partnership with a brother.

Andrew has an allotment of 400 acres at Katy. Last year he produced about 21½ barrels of rice to the acre.

He now has 10 tractors, two combines, three rice buggies and other tools and equipment for rice farming.

He has about $100,000 worth of equipment and buildings in Polk County and at Katy.

He and his brother have about 350 Brahman-Hereford cattle in Polk County.

Andrew believes in soil testing for needed chemical fertilizers. He is a member of the Katy Methodist Church, and a member of the Katy Young Farmers Chapter.

Reagan Brown, extension rural sociologist of Texas A. and M. College, was the principal speaker for the Jaycee dinner.

Dan Clinton, Jr., of 5631 Edith, chairman of the Junior Chamber of Commerce agriculture committee, presided.

Andrew will represent the Houston area at the state contest in Greenville Feb. 17.

Outstanding Farmer of the Houston Area – 1961

Award Highlight Newspaper Clipping

1969: 43 Years Old

Our property in Colorado County between Eagle Lake and Wharton was a beautiful place with about half of it wooded with pin oak, live oak, post oak, pecan, and some walnut trees. And with lots of brush and scrub oak. The lower half (about four hundred fifty acres) was the rich Colorado River red soil and had hundreds of pecan trees, of which there were three of the largest trees I have ever seen. It took three people with arms outstretched to reach around the circumference of the trunk. The branches reached out about eighty feet each direction and shaded about a half-acre of land.

One spring morning, I was there driving through the woods. I stopped to check the pecan trees to see if the little pecans were setting on, just wondering if there might be a good pecan crop that year. I walked up to the low hanging branches that were about head high. The grass was lush and green and was about knee high. I looked down at my feet and was startled to see some kind of animal lying right at my toes. It was a spotted baby fawn. I picked it up and put it in the pickup. I brought it home to Katy and we raised it on baby formula. It made the nicest pet; it would come in the house, go down the hallway to the girls' bedroom. It would run 'round and 'round the yard, which had a chain link fence, with the German Shepherd dog. It had no odor and seemed to be house broken from the very start.

1970: 44 Years Old

One year, I leased some property near Cat Springs, TX for deer hunting, along with several other men. It was close to home, and I could run over there some afternoons or weekends and come home the same night.

The property was nicely wooded and had lots of deer and had a nice creek running through it. The creek was about twenty-five feet wide and was real sandy with just a little stream of water. I was looking for a good place to hunt, and I went down into the sandy creek bottom to look for deer tracks. To my amazement, there were deer tracks everywhere, and I decided to make a deer blind on the ground and hunt right there on the creek bank. I made a blind of tree branches around three trees that were close together in a triangle. Next morning, I was anxious to hunt because there were so many deer signs and I had a good chance to get a shot at a buck. I got there before daylight and got in my blind to wait for sun-up.

In a little while, I heard splashing in the water in the creek. My heartbeat quickened; I thought sure deer were crossing right there, but I couldn't see them yet. As soon as it got light enough to see, I spotted another hunter in a blind on the other creek bank, just across from me. He had made the splashing noise, and I left right then as it would have been very dangerous.

Editor's Note: Any game shot or fish caught were used for food. Nothing was wasted.

I purchased a dozer to get the new place ready for rice. The upper 450 acres were Eagle Lake prairie land, nice and level with a good water supply, but it had scattered bunches of live oak and scrub oak. One of my hired men was operating the dozer and Dorothy and I were nearby. The dozer pushed out a small group of dead oaks and under the roots, about two feet deep, was a nest of Cottonmouth Moccasins. They are easily recognized; they are like no other snake. They are nearly black, short, three feet long with a stub tail, about two and a half inches in diameter, and have a cotton-white mouth. We killed them – five or six – with a shotgun I always carried in my pickup.

Then I walked over to where Dorothy was standing, and I looked down at her feet, and a huge Cottonmouth Moccasin just like the ones we had killed had crawled to within three inches of her heel. I was horrified, and I yelled to her "don't move, don't move." Then, with one swing, I clubbed it in the head because I didn't want to shoot that close to her foot.

Editor's Note: cottonmouth snakes are venomous vipers.

Certificate to Farm – 1977

Appendices

Obituary

Gene was born in Grand River, Iowa, on November 16th, 1926, to Lyman Irl and Myrtle Brown Andrew. At the age of ten, he and his family moved to Katy, TX, where they farmed rice in what is now the Mason Road and Cinco Ranch area. Gene attended Katy schools until the eleventh grade, at which time his family moved to Waller, TX. He graduated from

Waller High School in 1943. Gene enlisted in the Navy during WWII at the age of seventeen and proudly served his country for two years, being honorably discharged in 1946. He married Dorothy Knebel of Waller on December 31, 1954. They lived in Katy farming rice and raising their three children: Donna Gayle, Tammy Sue, and Myron Dale.

In 1962, Gene was honored as Outstanding Young Farmer by the Houston Junior Chamber of Commerce. Gene retired from farming in 1987, and he and Dorothy moved to Waller. Gene enjoyed gardening, traveling, riding his motorcycle, and raising a small herd of cattle. Gene and Dorothy especially enjoyed traveling with friends in the Northeastern U.S. to see the fall foliage.

Throughout his life, Gene enjoyed many sports – especially football and baseball. He was a lifelong supporter of Katy High School football, attending the first game ever played by Katy, and rarely missed a game up until his final illness.

One of his favorite pastimes in retirement was restoring old farm equipment. One of his last projects was a classic Farmall tractor that he painstakingly took apart piece by piece, cleaned, repaired, repainted, and rebuilt. He also had a knack for designing and building tools and implements from scratch for use around the farm. His ingenuity in this area was a unique and rare quality. Working with his hands with wood and metal were things he truly enjoyed.

Gene's character was sterling. He stressed honesty and trustworthiness and lived this in his daily walk throughout his lifetime. He was truly a man of his word and his handshake was as good as a signed contract. Gene was a devoted father and grandfather. Nothing was as important as, or more valued than, his family. He was a committed Christian and a founding member of the Katy Community Church, now Katy Bible Church. Gene loved his Savior, his country, and his family.

Beloved husband, father, and grandfather, Gene Russell Andrew, age 82, went to be with his Lord and Savior on May 19, 2009 after a lengthy illness.

Manuscript

Handwritten Page from Original
Manuscript

② probably pretty shy at first. My dad was so busy he just dropped Harold & I off at the front door the first day and we had to enrol ourselves. I was 10 yrs old and in the 7th grade. The superintendent and principal took one look at me and said "no way." They put me in the 5th grade and Harold in the 7th. I did poorly at first, I ranked 23rd out of a class of 24. By the fourth 6 weeks, I had moved up to #2 and the next one to #1. But a certain girl who had been at the top all year began crying so, Mrs Buening, my first teacher, changed her report card to #1 and mine to #2. This had to be my introduction to politics, however small.

Handwritten Page from Original Manuscript

ALSO BY JUNG WORKS

Finish the Race | The Great Mountain Adventure

Four cars embark on an exciting race through the mountains when they encounter unexpected challenges along the way. Who will win? Will they even be able to finish the race?

A fun children's book for ages 0-7. Written and illustrated by Thomas C. Jung.

Visit jungworks.com to see purchasing options in various formats (print, hardcopy, audiobook, ebook.)

Finish the Race | Super Cars Origin Story

The adventure continues! The race cars have just encountered a magic gas pump. Will they receive special powers?

A fun children's book for ages 0-7. Written and illustrated by Thomas C. Jung.

Visit jungworks.com to see purchasing options in various formats (print, hardcopy, audiobook, ebook.)

1926 | American Scenes

1926 | American Scenes is available in eBook, paperback, hardcover, and audiobook. To learn how to purchase, please visit jungworks.com.

Earlier Days Flying

Earlier Days Flying is available in eBook,
paperback, hardcover, and audiobook.
To learn how to purchase, please visit
jungworks.com.

FOUND A TYPO?

Found a typo? Email christine@jungworks.comwith the subject line of "Typo."

I've tried my very best to make sure this copy is perfect, but I'm only human! If you found an error, do let me know. I want to make sure the next reader has the very best possible version of this book. Thank you, in advance, for taking the time to submit an email to contribute to making this book even better!

BULK DISCOUNT

Interested in buying 10 or more copies of *1926*?

Contact christine@jungworks.com with "Bulk Discount" in the subject line.

ABOUT CHRISTINE JUNG

The editor and publisher of this book, Christine Jung, is Gene Andrew's granddaughter (his first grandchild.)

Christine is a writer, equestrian, classical pianist, and artist. Her days start with a cup of coffee and end with a glass of wine (preferably enjoyed outdoors.) In addition to being momager to a young actor, Christine keeps busy with her full-time job, six animals, and whatever her latest project might be. She is inspired by the books she reads, the places she travels and motivated by her supportive husband and son.

jungworks.com

About Jung Works

Jung Works is the umbrella for **Christine Jung, Thomas C. Jung**, & **Ethan Jung**'s projects. Three books, a screenplay, & new art are in the works! We invite you to join us on our creative journey.

Jung Works... A Creative Place
jungworks.com

www.ingramcontent.com/pod-product-compliance
Lightning Source LLC
Chambersburg PA
CBHW060834220526
45466CB00003B/1107